The Story of

CARBON

Mark D. Uehling

A FIRST BOOK

FRANKLIN WATTS
A Division of Grolier Publishing
New York London Hong Kong Sydney
Danbury, Connecticut

Photographs copyright ©: Bettmann: pp. 6, 25; Photo Researchers, Inc.: pp. 8 (Alfred Pasieka/SPL), 11, 20 (both Will & Deni McIntyre), 15 (Charlie Ott), 17 (Phil Farnes), 30, 31 (both Ken Eward/Biografx), 35, 42, 45 (all Ken Eward/SS), 47 (Jack Fields), 49 (Francoise Sauze/SPL), 50 (Ray Ellis), 53 (Sam Pierson, Jr.); Comstock Stock Photography: pp. 13, 28; Fundamental Photographs: p. 22 (Paul Silverman).

Library of Congress Cataloging-in-Publication Data

Uehling, Mark.
   The Story of carbon / by Mark D. Uehling.
       p.   cm.—(A First book)
   Includes bibliographical references (p.   ) and index.
   Summary: Discusses the chemical element carbon: its forms, uses, and importance in our lives.
   ISBN 0-531-20212-7
   1. Carbon—Juvenile literature. [1.Carbon.] I. Title. II. Series
QD181.C1U35   1995
546'.681—dc20   95-32186
                                                                        CIP AC

Contents

Chapter 1

A CINDERELLA STORY

Deep inside caves in southern France, there are drawings of horses, bison, and rhinoceroses on the cave walls. They are works of art from the Stone Age, and they may be some of the first drawings that people ever made. How did Stone Age people draw without pencils, paint, or crayons?

One of the first things that people ever used for drawing was charcoal. This material is the same as the black coals you may have seen burning in an outdoor grill. Charcoal is really just charred wood—wood that has been partially burned, or blackened, by fire. Stone Age people got their charcoal from their campfires.

It is easy to imagine a cave person picking up a piece of charred wood left from a fire and discovering that

*A cave artist drew these animals 30,000 years ago
using charcoal, a type of carbon.*

it would make marks on rocks. You may have done it yourself. Charred wood is soft, even softer than chalk.

This black material, this charcoal, is actually a substance called carbon. If Stone Age people had known all the wonderful surprises to come from carbon, they would have been amazed. They may have already noticed that burning charcoal makes less smoke than burning wood.

People would soon begin using charcoal instead of wood for their cooking and heating. It became an important fuel.

Later, people discovered carbon fuels buried underground that were even better than charcoal. They were coal and oil, which both contain mostly carbon. Today these two fuels provide the power for heating, lighting, and almost everything else in our world. Oil is so important to us now that people sometimes call it black gold.

The soot that rises into the air when carbon fuels burn is another kind of carbon. You may have seen this fine black powder inside chimneys or fireplaces. The Chinese, the Hindus, and the Greeks all learned in ancient times to make ink by mixing soot with plant oils. Even modern ink contains carbon that comes from soot. In fact, the words you are reading right now contain carbon. That's why they're so black.

Carbon has played a big part in drawing and writing. There is carbon in pencils as well as in ink. The gray-colored lead of a pencil is a form of carbon called *graphite*. It is soft, but it doesn't smudge as much as charcoal does.

Until the last century, people did not realize how much carbon had done for them. They were grateful for fuels, ink, and graphite, but they didn't know that they all contain carbon. Carbon's many talents were not appreciated. That finally began to change when scientists learned how to recognize carbon. When they did, they discovered something astounding. It turned out that diamonds are actually charcoal in disguise!

*Carbon sometimes appears in the form of
a diamond—as dazzling as Cinderella at the ball.*

Even scientists found it hard to believe that clear, sparkling gems could be the same as soft, black powder. Then scientists began to see a whole other side of charcoal, or carbon. It was a lot like what happened when Cinderella put on the glass slipper. At last people could see her true nature. They realized she could do much more than clean cinders from the fireplace.

As scientists got to know carbon better, they discovered many more wonderful things about it. They learned that it is vital for life on earth. It is a big part of every living thing.

As you read on, you will get to know carbon, too. You will find that it has done many wonderful things for us and our planet. But you will also discover that if it is not used wisely, it can cause problems. By the end of this book, you will have learned much of what scientists know about carbon and all the ways it has affected our lives. This is the story of carbon.

Chapter 2

FUELING CIVILIZATION

People first got to know carbon as charcoal. Stone Age people liked charcoal because it not only burned with less smoke but it also gave off more heat than wood did.

So charcoal became popular as a fuel with Stone Age people. They made campfires by piling charcoal in the center of a ring of large stones. The stones kept the fire from spreading.

One day, scientists believe, someone tending a charcoal fire must have made an amazing discovery. Maybe it happened when a green stone in the circle came up against a piece of burning charcoal. Perhaps as the stone heated up, the lucky cave person noticed drops of a

Charcoal burns hotter and gives off less smoke than wood does.

shiny orange liquid coming from it. When the liquid cooled, it became a solid metal the color of a penny.

This metal was copper. The green stone was copper *ore*, a rocky material from underground. Early humans learned to make tools, weapons, and jewelry out of copper. While it was liquid, copper could be easily formed into any shape they wanted.

If it were not for charcoal, copper would not have come from the green rock. Wood does not burn hot enough to separate copper from its ore. But charcoal does. People later found that charcoal could bring silver, iron, and other metals out of other kinds of rocks.

Heating rocks to draw out metals was some of the first chemistry people were involved with. This process is called *smelting*. The Egyptians and Mesopotamians were doing it 6,000 years ago—in 4000 B.C. They didn't know why it worked; they may have even thought it was magic. Smelting was so mysterious and so important to early humans that they gave priests the responsibility for it. These priests might be considered the world's first *chemists*. Chemists are scientists who study the materials that make up everything in the world.

Smelting helped humans advance by giving them tools to work with. Iron, in particular, was so important to human progress that we call the period after its discovery the Iron Age. Beginning about 1200 B.C., iron allowed people to make much stronger tools and weapons.

Smelting iron in very hot charcoal made it even stronger. People didn't know it then, but the iron was

Steel frames support the two towers of the World Trade Center
in New York City. Carbon in steel gives it
the strength to support these and other skyscrapers.

absorbing carbon. Today, industries add carbon to iron to make steel. Steel is a material used in bridges and sky-scrapers. Skyscrapers would not be possible without the strength that carbon gives to steel.

In ancient times, charcoal became so popular for smelting and heating that many trees were cut down to make it. Large sections of forest disappeared in North Africa. Fortunately, other fuels eventually were discovered underground.

Perhaps the first clue that there could be fuel under-ground came as long ago as 6000 B.C. People in the Middle East noticed a *gas* rising out of the ground. This gas was invisible, like air, but they could smell it. Middle Easterners found that they could burn the gas to make a fire. Persians who lived in the Middle East in ancient times worshiped fire by burning the gas in their religious cere-monies. Today this gas is called *natural gas*.

These ancient people probably noticed black oil seeping up from underground near the gas holes. Oil and natural gas often come out of the ground together. When the black liquid was set on fire, it burned as the gas did. Ancient Egyptians learned to use the oil in medicines, and ancient Persians used it to make fire bombs to shoot at their enemies.

Another kind of underground fuel looks a lot like charcoal. It even has a similar name—coal. The ancient Chinese were using coal to smelt copper in 1000 B.C. But Europeans did not discover it until much, much later.

Early humans may have suspected that charcoal, coal, and oil are related somehow. After all, they are all black and they all create fire. But there was no way for them to know that these fuels, as well as natural gas, were made mostly of carbon.

At that time, people had no idea what materials were made of. A few hundred years B.C., the ancient

You can see black stripes of coal in these rocks in Alaska.

Greeks guessed that some basic substances combine to make everything in the world. They called these substances *elements*.

One Greek scientist, named Empedocles, thought there were four elements. He believed that all *matter*, anything that takes up space, is a combination of earth, air, fire, and water. Even though air is invisible, Empedocles knew that it is a kind of matter. He had seen a bubble of air take up space in water.

Most people believed Empedocles's idea for 2,000 years. This belief lasted so long partly because many early scientists thought it explained what happens when wood burns. To them, it seemed that wood was breaking down into its elements as it burned.

First of all, the fire appeared to come out of the wood. And the smoke that rises from the wood was a lot like air. It even disappeared into the air. They could also sometimes see water oozing out of burning wood. The ashes left behind after a fire looked like earth, which is really just dirt. So early scientists thought wood was made of earth, air, fire, and water.

If only the Greeks knew what we know today. The main element in wood is actually carbon! It was carbon in wood, charcoal, and other fuels that provided heat and light all those years. And they became even more important later on. Without heat and light, civilization could not have developed the way it did.

Civilization began when people started gathering together in towns and developing farming and industry.

Burning wood produces ashes, smoke, fire, and sometimes drops
of water. This may be where the ancient Greeks got the idea
of four elements—earth, air, fire, and water.

The smelting of metals in carbon fuels was the beginning of industry. When people learned to make carbon steel in the nineteenth century, our civilization came into the Industrial Age. Carbon is used for many other purposes in modern industry, such as for coloring things black. In fact, if you see something black, such as a car tire or a record album, you can bet it has carbon in it.

So you see, carbon has helped advance civilization in several ways. Reading and writing made it easier for people to share ideas with each other. The ancient Greeks write with carbon ink and the form of carbon called graphite. In fact, *graphite* comes from a Greek word meaning "to write."

Carbon fueled civilization for thousands of years before anyone really knew what it was. Like Cinderella slaving away for her stepmother and stepsisters, carbon quietly served humans with no recognition. Before carbon could be given the recognition it deserved, a prince of a chemist named Antoine Lavoisier would have to discover the true elements.

Chapter 3

DISCOVERING CARBON

One of the first people to question the Greek idea of the elements was a man named Jan van Helmont. He lived in Belgium around 1600. He believed that all matter, except for air, is made of water. It turned out that he was wrong about that. But he was right about one thing. He knew that fire could not be an element because it is not a substance. It merely changes a material into something else.

So when Helmont watched charcoal burn, he realized that the fire was changing it into another substance. But he wondered why burning charcoal produced only a small amount of ashes. When he started with 62 pounds of charcoal, he ended up with only 1 pound of ashes. Where did the missing 61 pounds go?

Experiments with burning charcoal were the first steps toward discovering carbon.

Helmont knew that matter could not just disappear. Most chemists before him didn't think it was important to weigh the materials left after mixing or burning. But fortunately for carbon, Helmont started weighing everything. His experiment to find out what happened to the 61 pounds of charcoal was the first step toward discovering carbon.

In this experiment, he burned the charcoal in a closed glass container. He wanted to trap anything that might escape from the charcoal. To his surprise, the glass exploded! Some very powerful substance had indeed escaped from the charcoal. Since it was able to break through the glass, he called it a wild, invisible spirit.

He came across similar substances in other experiments. Even though most of them were invisible like air, Helmont realized that they were completely different from air. So he named them gases. Other chemists had noticed them before, but thought they were just different kinds of air.

A century later a Scottish chemist named Joseph Black proved that Helmont was right. Black was able to capture the gas from burning charcoal in a bottle. To see whether the gas was like air, he tried several experiments. First, he put a burning candle into the gas. The flame immediately went out. Then he wondered whether animals could live in it. When he put birds and other small animals in a container filled with the gas, they died. This gas most certainly was different from air!

Black discovered one other very interesting thing about the gas. It is the same gas that people and animals breathe out, or exhale. Today this gas is called *carbon dioxide*, but Black never found out exactly what it was.

By 1772, an English chemist named Joseph Priestley found some more clues about this mysterious gas. He happened to live next door to a brewery, a place where beer is made by fermenting yeast. When yeast ferments, it gives off carbon dioxide, the same gas that rises out of burning charcoal. If you look at beer in a glass or bottle, you will probably see tiny bubbles of carbon dioxide rising in the beer. The same gas makes bread rise. The brewery made so much carbon dioxide that Priestley was allowed to study some of it.

One day Priestley tried mixing carbon dioxide with water. It rose through the water in tiny bubbles as it did in the beer. He liked the way the bubbly water tasted so he

Carbonated drinks have bubbles of carbon dioxide gas added to them.

suggested that the brewery sell it. He had invented what we call *soda water*. Eventually other people would use carbon dioxide to make bubbly soft drinks, such as Coca-Cola and Pepsi. These drinks are said to be *carbonated*. But Priestley's most amazing discovery was yet to come.

Priestley was fascinated by gases. From his experiments, he was beginning to see that there are many kinds. One kind of gas made flames burn larger and was particularly good for breathing. He didn't know it then, but this gas was *oxygen*. It seemed to be the opposite of carbon dioxide. Mice did very well in oxygen, but they died when there was only carbon dioxide to breathe.

He wondered whether plants would die in carbon dioxide too. To his surprise, a plant grew even better in a container of this gas than in air. Then he tried putting a mouse into the container with the plant. Now the mouse was able to live!

Priestley realized then that without even knowing it, plants and animals cooperate to help each other live. Animals inhale oxygen and exhale carbon dioxide. Plants take in carbon dioxide and give off oxygen through tiny holes in stems and leaves. But Priestley had no idea that carbon was an important part of this process.

Priestley's experiments convinced more and more scientists that air was not an element. If air was made of oxygen, carbon dioxide, and other gases, how could it be an element? Scientists had already begun having doubts about fire and earth. But most of them still believed that water was an element.

One, however, was determined to find out whether that was true. He was the French chemist Antoine Lavoisier. Like Priestley, he began experimenting with many of the new gases, particularly oxygen. He thought they might hold the secret of the true elements.

About 1778, he made a surprising discovery: the gas from burning charcoal is actually a combination of charcoal and oxygen from the air. Even today, this seems incredible. How could a solid black substance like charcoal become an invisible gas by combining with oxygen?

When charcoal burns, a *chemical reaction* is taking place. Oxygen is combining chemically with charcoal. The two materials are not just mixing together like gases in air or like ingredients in a recipe. You can usually see or taste the ingredients of a mixture. But the result of a chemical reaction can be very different from what you started with.

Lavoisier was the first scientist to realize that whenever anything burns in air, it is combining chemically with oxygen. Water, he soon discovered, is the result of a similar reaction. It is the combination of oxygen and another invisible gas called *hydrogen*. When hydrogen is set on fire, water forms. Lavoisier proved that water is not an element. It is a *compound* of the elements oxygen and hydrogen. Likewise, carbon dioxide is a compound of the elements carbon and oxygen.

Lavoisier gave these new elements their names. He called charcoal *carbone*, which is the Latin word for charcoal. He named more than twenty other substances he

Antoine Lavoisier recognized that carbon is an element and gave it its name. His wife helped him carry out many of his experiments.

thought were elements. Many of them were metals that had been known since ancient times, including copper, iron, and gold. He turned out to be right about most of them.

It soon became clear what is going on when metals are smelted. A compound is being separated into its elements in a chemical reaction. The smelting of iron ore is a good example. Iron ore is a rocky material that is a compound of iron and oxygen. When iron ore is heated with charcoal, oxygen rises into the air and melted iron flows out.

Lavoisier began giving compounds chemical names that reveal what elements they contain. Iron ore eventually became known as *iron oxide*. An *oxide* is a compound containing oxygen.

Lavoisier's discoveries started a whole new way of thinking in chemistry. His ideas were so different that some people would not accept them at first. But gradually, Lavoisier convinced chemists to throw out the theory of the four elements and start searching for the true elements. Over the next 200 years, chemists discovered a total of 92 elements in nature. Lavoisier's work in 1789 was really the start of chemistry as we know it today.

And it was the start of an exciting time for carbon. With the help of Priestley and the others who had come before him, Lavoisier was beginning to see that carbon was more than charcoal. He was realizing that carbon is very important for living things. And he was just beginning to suspect that it was somehow related to diamonds.

A DAZZLING DISCOVERY

At one time, Lavoisier thought clear, sparkling diamonds were made of water. But he realized that was wrong when he and some other scientists set diamonds on fire in the

1770s. It may sound unbelievable that something as hard as a diamond could burn. People at the time were surprised too.

The scientists used an enormous magnifying glass to focus the sun's heat on a small diamond. As the diamond burned, they discovered, it gave off carbon dioxide gas as charcoal did. They knew that meant diamonds had to have carbon in them.

But the scientists thought that diamonds must be a compound containing carbon. They didn't realize that diamonds *are* carbon. They must have thought it was impossible for soft, black charcoal to be the same material as a clear, hard diamond.

Even after an English chemist, Smithson Tennant, proved in 1796 that a diamond is all carbon, some scientists refused to believe it. Tennant burned the same weight of charcoal and diamonds, and found that the same amount of carbon dioxide came from each material. That could only mean that diamonds and charcoal were the same material.

About that time, a Swedish chemist named Carl Scheele did a similar test with graphite. It proved that graphite is carbon too. Before that, people had often confused graphite with the metal lead because they are both gray in color. That is why people still say "lead pencils" today. The "lead" of a pencil is really graphite.

How could charcoal, graphite, and diamonds all be the same material? Chemists would not find out until the next century. First, they had to learn more about what

*Even scientists could not believe that diamonds are made
of the same material as charcoal and graphite.*

makes one element different from another. Why, for
instance, does carbon look and behave so much different-
ly than oxygen?

It was an English chemist, John Dalton, who showed other chemists the answer to that question. Early in the nineteenth century he was studying how elements combine to form compounds. He and other chemists had noticed that each compound always contains the same amount of each element. For every pound of hydrogen in water, there are always 8 pounds of oxygen. For every 12 pounds of carbon in carbon dioxide, there are always 16 pounds of oxygen.

The reason that happens, Dalton guessed, is that each element must be made up of many tiny particles called *atoms*. When elements form compounds, their atoms bond together in groups called *molecules*. There is always the same number of atoms from each element in a compound's molecules. A molecule of carbon dioxide, for example, always contains one carbon atom bonded to two oxygen atoms. That's why there is always the same ratio of carbon to oxygen.

Although these atoms are much too small to see, tests eventually proved that they do exist. Chemists began representing compounds by formulas that showed which atoms were in each molecule. They abbreviated carbon as C and oxygen as O. The formula for carbon dioxide is CO_2 because there are two atoms of oxygen. The *di* in carbon dioxide also means it has two oxygen atoms.

It is molecules, Dalton realized, that make compounds look and behave the way they do. Molecules give compounds their *properties*, a chemist would say. And it is differences in atoms that give elements their properties.

But the way atoms are arranged in an element also determines its properties. In charcoal, carbon atoms are jumbled together like bubble gum balls in a machine at the grocery store. But in graphite, the atoms line up neatly in layers. That's why it's so easy to write with. Every time you make a mark with a pencil, several layers of carbon atoms slide off the graphite.

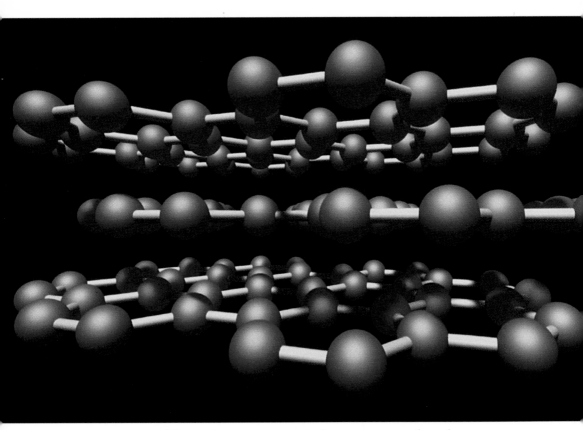

The carbon atoms in graphite are arranged in sheets.

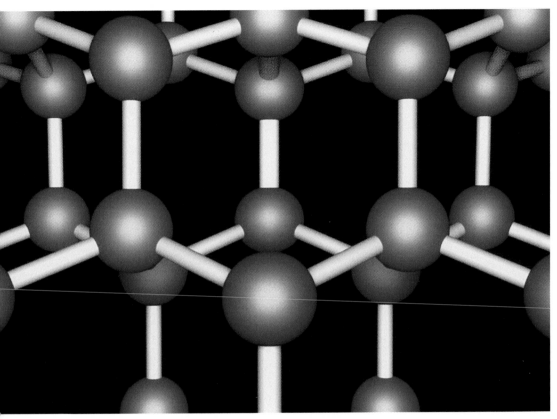

The carbon atoms in diamonds are arranged in stiff pyramid shapes.

The atoms in a diamond are in a much more rigid pattern than those in graphite. Four atoms arranged in the shape of a pyramid are repeated over and over again throughout a diamond. This pattern is so stiff that it makes diamond the hardest material found naturally on earth. Also because of the pattern, light bounces around inside diamonds, making them glitter.

Diamonds are a kind of *crystal*. A crystal is a solid material with atoms arranged in a pattern. Graphite is a kind of crystal, too, even though it doesn't sparkle as most crystals do.

Most elements don't have as many disguises as carbon does. Chemists call different forms of the same element *allotropes*. Carbon's allotropes are unusual because they are so different from one another. That's one reason it took so long for scientists to recognize all its capabilities. They found that one of carbon's best qualities is its variety.

So you see, by the end of the eighteenth century carbon's prince had come. You might think of him as Lavoisier and all the other chemists who explored and searched until they found all the places carbon was hiding. In the next century, that search would continue into the chemicals of life.

Chapter 4

THE SPICE OF LIFE

Not long after Lavoisier discovered the elements, chemists began finding carbon in more and more compounds. Lavoisier himself had noticed that sugar, alcohol, and plant oils all contain carbon. These foods are compounds that come from plants or animals. Because the compounds are found in living organisms, chemists called them *organic*. They soon discovered that it was carbon that gave organic compounds their great variety.

Before long, chemists discovered so many organic compounds that a whole branch of chemistry was created for studying them. It was Lavoisier who really started it all by finding a way to figure out which elements were in an organic compound.

When Lavoisier burned sugar, alcohol, or oil, all that remained was carbon dioxide and water. That told him that the original compounds must contain carbon and hydrogen and maybe oxygen. Burning the compounds simply added some oxygen and rearranged the atoms to form carbon dioxide and water.

After each compound burned completely, Lavoisier weighed the carbon dioxide and the water. Knowing these weights, he could figure out how many carbon, hydrogen, and oxygen atoms were in each molecule of the compound.

Other chemists learned this trick and improved it. They began burning all kinds of organic substances and figuring out what they were made of. They were surprised at all the ways carbon could combine with other elements. After a while, chemists became so confused that they began organizing carbon compounds into groups. They put compounds with similar formulas in each group.

One group was compounds containing only carbon and hydrogen. They are called *hydrocarbons*, and there are thousands of them. The simplest has a molecule with one carbon atom and four hydrogen atoms—CH_4. It is a gas called *methane*, and it is the main ingredient in natural gas. Another hydrocarbon has 40 carbon atoms and 56 hydrogen atoms. $C_{40}H_{56}$ gives the red color to tomatoes and watermelons. Coal, oils from plants, and oils from underground are also made of hydrocarbons.

Sugars are in another group. The sugar you put on

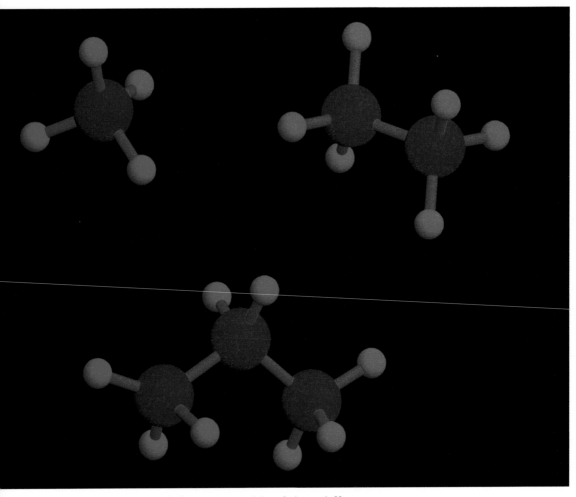

These are models of three different
hydrocarbon molecules—methane on the upper left,
ethane on the upper right, and propane below.
The carbon atoms are dark blue and
the hydrogen atoms are aqua.

your cereal is a compound called *sucrose* that comes from sugarcane. Its formula is $C_{12}H_{22}O_{11}$. The sugar in your blood is *glucose*—$C_6H_{12}O_6$. The sugar in fruit is *fructose*. And there are many others. But in every sugar there are half as many oxygen atoms as hydrogen atoms.

Alcohols too are in their own group. Their molecules always have only one oxygen atom. There are many different kinds of alcohols, but the kind people drink is *ethanol*—C_2H_5OH.

As the first organic chemists studied the formulas in each group, they began to see a pattern to the way carbon bonds with other elements. The discovery of this pattern helped solve the mystery of how atoms bond to each other.

FAMILY TIES

What makes atoms hold together so tightly? It has to do with tiny particles inside atoms. There are *protons* and *neutrons* in the center of the atom, which is called the *nucleus*. *Electrons* are much smaller particles that fly around the nucleus. You might think of them as flies buzzing around an orange.

The simplest atom is the hydrogen atom. It has only one proton and one electron. A carbon atom has six protons and six electrons. The number of electrons in an atom is always the same as the number of protons. Oxygen, for example, has eight protons and eight elec-

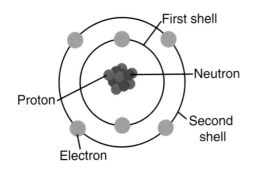

First shell

Neutron

Proton

Second shell

Electron

A carbon atom has six protons and
six neutrons in its nucleus, with six electrons
circling it in two shells.

trons. The number of neutrons in a carbon atom varies, but the most common form of carbon has six neutrons. An element's properties depends on the number of protons, neutrons, and electrons in the atom.

One of carbon's properties is its friendliness. It likes to bond with other atoms, especially hydrogen, oxygen, and other carbon atoms. The reason for this is the way carbon's electrons are arranged. Electrons like to gather together in families. Two of carbon's six electrons form a family close to the nucleus. Chemists think of them as living in a *shell* that can hold only two electrons.

The remaining four electrons start another family farther away from the nucleus. But this second shell can hold eight electrons, so it has four empty spots for electrons. Atoms, it seems, are not happy unless the families in their outer shells are complete. That's why they bond with

Periodic Table

1 H 1.00794 Hydrogen								
3 Li 6.941 Lithium	**4** Be 9.01218 Beryllium							
11 Na 22.98977 Sodium	**12** Mg 24.305 Magnesium							
19 K 39.0983 Potassium	**20** Ca 40.078 Calcium	**21** Sc 44.95591 Scandium	**22** Ti 47.88 Titanium	**23** V 50.9415 Vanadium	**24** Cr 51.9161 Chromium	**25** Mn 54.93805 Manganese	**26** Fe 55.847 Iron	**27** Co 58.9332 Cobalt
37 Rb 85.4678 Rubidium	**38** Sr 87.62 Strontium	**39** Y 88.9059 Yttrium	**40** Zr 91.224 Zirconium	**41** Nb 92.9064 Niobium	**42** Mo 95.94 Molybdenum	**43** Tc (98) Technetium	**44** Ru 101.07 Ruthenium	**45** Rh 102.9055 Rhodium
55 Cs 132.9054 Cesium	**56** Ba 137.327 Barium	**57** La 138.9055 Lanthanum	**72** Hf 178.49 Hafnium	**73** Ta 180.9479 Tantalum	**74** W 183.85 Tungsten	**75** Re 186.207 Rhenium	**76** Os 190.2 Osmium	**77** Ir 192.22 Iridium
87 Fr (223) Francium	**88** Ra 226.025 Radium	**89** Ac (227) Actinium	**104** Unq (261) (Unnilquadium)	**105** Unp (262) (Unnilpentium)	**106** Unh (263) (Unnilhexium)	**107** Uns (262) (Unnilseptium)	**108** Uno (265) (Unniloctium)	**109** Une (266) (Unnilnonium)

In the periodic table, chemists have organized the elements according to atomic number, the number of protons in an atom of the element. The elements in any column have the same number of electrons in their outermost shells. Carbon is versatile because it is in a column of elements whose outer shell is half filled.

58 Ce 140.115 Cerium	59 Pr 140.9077 Praseodymium	60 Nd 144.24 Neodymium	61 Pm (145) Promethium	62 Sm 150.36 Samarium
90 Th 232.0381 Thorium	**91** Pa 231.0359 Protactinium	**92** U 238.029 Uranium	**93** Np 237.048 Neptunium	**94** Pu (244) Plutonium

of the Elements

CHEMICAL SYMBOL

ATOMIC NUMBER

ATOMIC WEIGHT

ELEMENT NAME

					2 He 4.00260 Helium
5 B 10.811 Boron	**6** C 12.011 Carbon	**7** N 14.067 Nitrogen	**8** O 15.994 Oxygen	**9** F 18.998403 Florine	**10** Ne 20.1797 Neon
13 Al 26.96154 Aluminum	**14** Si 28.0855 Silicon	**15** P 30.973762 Phosphorous	**16** S 32.066 Sulfur	**17** Cl 35.4527 Chlorine	**18** Ar 39.948 Argon

28 Ni 58.693 Nickel	**29** Cu 63.546 Copper	**30** Zn 65.39 Zinc	**31** Ga 69.723 Gallium	**32** Ge 72.61 Germanium	**33** As 72.9216 Arsenic	**34** Se 78.96 Selenium	**35** Br 79.904 Bromine	**36** Kr 83.80 Krypton
46 Pd 106.42 Palladium	**47** Ag 107.8682 Silver	**48** Cd 112.41 Cadmium	**49** In 114.82 Indium	**50** Sn 118.71 Tin	**51** Sb 121.757 Antimony	**52** Te 127.60 Tellurium	**53** I 126.9045 Iodine	**54** Xe 131.29 Xenon
78 Pt 195.08 Platinum	**79** Au 196.9665 Gold	**80** Hg 200.59 Mercury	**81** Ti 204.383 Thallium	**82** Pb 207.2 Lead	**83** Bi 208.9804 Bismuth	**84** Po (209) Polonium	**85** At (210) Astatine	**86** Rn (222) Radon

63 Eu 151.965 Europium	**64** Gd 157.25 Gadolinium	**65** Tb 158.9253 Terbium	**66** Dy 162.50 Dysprosium	**67** Ho 164.9303 Holmium	**68** Er 167.26 Erbium	**69** Tm 168.9342 Thulium	**70** Yb 173.04 Ytterbium	**71** Lu 174.967 Lutetium
95 Am (243) Americium	**96** Cm (247) Berkelium	**97** Bk (247) Berkelium	**98** Cf (251) Californium	**99** Es (252) Einsteinium	**100** Fm (257) Fermium	**101** Md (258) Mendelevium	**102** No (259) Nobelium	**103** Lr (260) Lawrencium

other atoms. The atoms help each other complete their families by sharing electrons.

A carbon atom bonds with atoms that can share four electrons with it. A hydrogen atom has one electron to share. So four hydrogen atoms can complete carbon's outer shell. That is why a carbon atom bonds with four hydrogen atoms in methane. In this bonding, the carbon atom shares each of its outer electrons with a hydrogen atom. That helps each hydrogen atom complete its family of two electrons.

Carbon helps oxygen in the same way. Oxygen has two electrons in its first shell and six electrons in its outer shell. It needs two more electrons to be complete. One carbon atom, with four electrons to share, can satisfy two oxygen atoms. That's how carbon dioxide forms.

The bonds carbon atoms make with each other are the most interesting of all. In the nineteenth century, chemists began to realize this when they tried to figure out how some organic molecules could contain so many carbon atoms. They tried to imagine what these molecules look like. A German chemist named August Kekule was so fascinated by carbon atoms that he dreamed about them. One day as he napped on a bus, he dreamed of carbon atoms linking together in long chains. When he woke, he knew that this must be how the molecules were built.

It turned out that he was right. In these chains a carbon atom may share one, two, or three electrons with another carbon atom. The more electrons they share, the

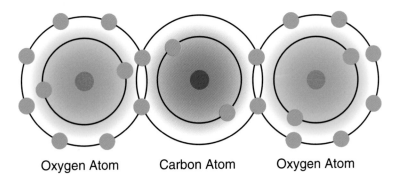

Oxygen Atom   Carbon Atom   Oxygen Atom

*A carbon atom bonds with two oxygen atoms to form
a carbon dioxide molecule. Atoms bond by sharing electrons
with each other.*

stronger the bond. Hydrogen or other atoms then fill in any empty spots along the chain.

Another dream showed Kekule the shape of other carbon compounds such as *benzene*. Benzene is a clear liquid that can be burned as a fuel. It is a hydrocarbon with six carbon atoms and six hydrogen atoms. For a long time, chemists couldn't understand how these atoms could share electrons to complete their families.

In Kekule's dream, he saw chains of carbon atoms that looked like snakes. The snakes were sliding around each other. All of a sudden, one of the snakes curled up and grabbed its own tail in its mouth. Kekule realized then that the carbon atoms in benzene could bond together in a ring. He was right about that too.

A molecule of benzene has a ring of six carbon atoms with a single hydrogen atom bonded to each one.

The idea that molecules could take such unusual shapes was surprising back then. The compounds that don't contain carbon have very simple shapes. Chemists soon discovered that carbon could create even more fantastic shapes. Every protein in your body has a different shape because of carbon. These shapes help proteins do their jobs and keep you healthy.

To form unusual shapes, the bonds between carbon atoms have to be very strong. Double and triple carbon bonds are among the strongest in the world. In a double bond, a carbon atom shares two electrons with another carbon atom. In a triple bond, each atom shares three.

Carbon can share its electrons in many ways because its second shell is half full of electrons. Without carbon, the variety of living organisms probably would not be possible. Its versatility is really what gives carbon a Cinderella quality.

Because of its versatility, carbon is in more compounds than all the other elements combined. There may be as many as two million carbon compounds. Without a doubt, carbon is the friendliest of all the elements.

COPYING NATURE

In the nineteenth century, organic chemists learned to make organic compounds from simpler chemicals. These chemicals did not come from living things. Even though the man-made compounds weren't really organic, they were still considered part of organic chemistry. Nowadays, organic chemistry means the study of carbon compounds, whether they come from organisms or not.

In this century, organic chemists started making new compounds that had never been seen before. Copying nature, they made long chains of carbon atoms and invented plastics. Today plastics are everywhere. They make things cheaper and lighter. Nylon, polyester, Teflon, and Styrofoam are also man-made organic chemicals. A molecule of Styrofoam has the same shape as the benzene ring.

In the last decade, scientists invented a completely new shape for carbon atoms. They found a way of connecting 60 carbon atoms into the shape of a ball. It was like a soccer ball, with many many flat triangles on its surface. These balls of carbon are called *buckyballs*. It is yet another allotrope of carbon.

Buckyballs are so unusual that many scientists have begun studying them. This material has some remarkable properties that may be useful to people. Recently, scientists discovered that nature invented buckyballs first. They found buckyballs in *meteorites*. Meteorites are large rocks that fly to Earth from outer space. The carbon in meteorites may have helped start life on Earth.

Another place buckyballs have been found is in soot! Carbon is truly a Cinderella element. The carbon that has dazzled scientists in the last few years is some of

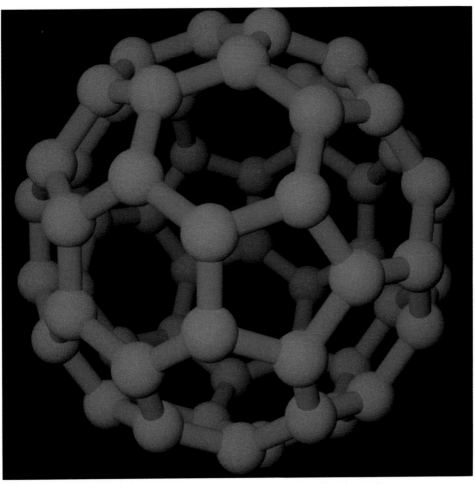

Carbon atoms join together in a ball in buckyballs.

the same carbon known in ancient times. In the next chapter, you will find another connection between the carbon in fuels and the carbon in living organisms.

Chapter 5

THE CARBON CYCLE

Before the nineteenth century, most lamps gave off light by burning oil from whales. There was no electricity then. You couldn't flip a switch to turn on a light. Back then, it was hard to get carbon fuels, such as coal, petroleum, and natural gas, to use for lighting, heating, and cooking. There were plenty of them, but it was too expensive to get them from the ground.

That all changed with the Industrial Revolution in the nineteenth century. Industries now needed lots of coal and petroleum to fuel furnaces and engines. So more mines and wells opened to bring the fuels up from underground. The fuels became cheaper and easier to get.

This tower is an oil well, which drills deep into the ground until it hits oil. Natural gas comes up from the well too, and if it is not collected, it is burned.

People began using fuel made from coal to light their lamps. They stopped killing whales for oil, which was a good thing, for whale hunting had greatly reduced the numbers of whales. People also began using carbon fuels to heat homes and other buildings. These fuels replaced wood and charcoal. Fortunately, that reduced the number of trees that were cut down. Carbon fuels seemed to be good for the environment.

As more carbon fuels were mined in the nineteenth century, scientists got clues to how the fuels formed in the earth. The remains of plants and animals that died long ago were often found with coal and oil underground. About this time, chemists were noticing how similar underground fuels are to compounds from living things. The molecules in fuels all contain carbon and hydrogen atoms. They are hydrocarbons, just like oil from your skin. Was the similarity just a coincidence?

Scientists eventually realized that underground fuels form over millions of years from the remains of plants and animals. These remains are called *fossils*. That's why coal, petroleum, and natural gas are called fossil fuels.

Here's what scientists think happened. When plant life began on Earth billions of years ago, the "air" was mainly carbon dioxide. There wasn't much oxygen in it. But of course, plants were very happy in all that carbon dioxide. They grew large and spread all over. Earth must have been one giant garden. After a while, all the oxygen the plants gave off made it possible for animals to live.

For heat, the Irish still burn logs of peat cut from bogs in Ireland.

But over time, great numbers of plants died and piled up on the ground. Before they could completely rot away, they were buried underground. First they became *peat*, a mosslike material that gardeners often mix with soil. You can find peat underground today in many places, including Georgia and Michigan. Before coal became plentiful, many people heated their homes by burning peat.

When peat is squeezed by rocks and heated underground, it turns into coal over time. Water and carbon dioxide escape from the peat. The peat becomes blacker and blacker as the concentration of carbon becomes greater. If the pressure continues on coal for a long time, it can turn into pure carbon. It can be turned into graphite.

Diamonds form when carbon in melted rock comes under very great pressure. But this doesn't happen very often. That's why diamonds are so rare and expensive. Most diamonds are found underground in Africa.

When the carbon in peat becomes concentrated, coal forms.

Scientists believe that petroleum, or oil, comes from plants and animals that die in the ocean. They sink to the bottom of the ocean and pile up there in layers upon layers. If they are buried under a lot of rock, the pressure and heat over many years may change them to oil.

Almost all the coal and oil under the ground formed from 345 million to 280 million years ago. Because a lot of carbon was concentrated in the earth then, scientists call that time the Carboniferous Period.

This process continues today, though not as much as back then. It is part of what is called the *carbon cycle*. In this cycle, carbon moves from the atmosphere through plants and animals and into the earth. Then it returns to the atmosphere.

At the start of the cycle, plants take in carbon dioxide from the air. The sun causes a chemical reaction in the plant called *photosynthesis*. Photosynthesis changes carbon dioxide and water into oxygen and *carbohydrates*. When animals eat plants, the carbohydrates change back into carbon dioxide and water. This chemical reaction releases energy that animals need to live.

Some of the carbohydrates are stored in animals too. When plants and animals die, the carbon in their bodies becomes part of fossil fuels. This decaying process releases some of the carbon back into the atmosphere as carbon dioxide. And when fossil fuels burn, more carbon goes into the atmosphere as carbon dioxide. Carbon has come full circle. In this cycle, carbon connects the earth with the plants and animals and with the atmosphere.

HOW WE AFFECT THE CARBON CYCLE

In our century, people have used more and more fossil fuels. Coal and oil make electricity and heat our homes. Gasoline, which is made from oil, makes cars run. Oil has become so important in the United States that there was a crisis in the 1970s because of it. Countries in the Middle East refused to sell oil to the United States. Americans began to worry that there would not be enough energy for everyone. Gasoline prices and electricity bills started rising fast.

This period was called the energy crisis. That's when people realized that fossil fuels would not last forever. The Earth is not making them nearly as fast as people are using them. Suddenly, carbon fuels became almost as precious as diamonds. People became more grateful for all the energy they provided over the years. Many people and governments began to look for ways to use less of these fuels.

People who are concerned about the environment have another reason to stop using so many fossil fuels. They pollute the air. First of all, burning coal gives off acids into the air. Then rain carries the acids into lakes and streams. The acids kill fish and other animals that live in the water. Animals that get their food from the lakes and streams may die too.

The carbon dioxide that all fossil fuels give off when they burn is another pollutant. Cars, factories, and power plants release a lot of carbon dioxide into the air. If

Coal-burning power plants give off carbon dioxide
and other pollutants.

too much of this gas builds up in the atmosphere, it can cause problems. Carbon dioxide in the air absorbs heat that comes from the earth. If the carbon dioxide weren't there, that heat would go into outer space. So all of this gas in the atmosphere acts like a blanket around the earth. Scientists call this the *greenhouse effect*.

Some scientists are afraid that too much carbon dioxide will heat up the atmosphere. Even if it heats up only a degree or two, a lot of ice at the North and South poles would melt. If that happened, the oceans would rise and land along the shores would flood.

To make things worse, there are fewer and fewer plants to absorb the carbon dioxide in the atmosphere. People are clearing large areas of forests to farm the land or build buildings. Scientists have found that the amount of carbon dioxide in the atmosphere has been rising over the last hundred years. But no one can be sure exactly what is causing it.

Many scientists think that the workings of the planet are so complicated that no one can predict whether the oceans will rise. But to be safe, most scientists say, we should burn less carbon fuels. In 1990, 170 scientists from all over the world asked industrial nations to promise to burn less carbon. Many of the countries made promises, but they have not yet been able to replace carbon fuels. Nuclear power could replace them, but it has problems too. Solar energy, wind energy, and other kinds of energy are too expensive right now. The best thing people can do is use as little energy as possible.

So the world is realizing that carbon should be treated with more respect. For many centuries, people took it for granted. There seemed to be plenty of it, so we burned lots of it. We didn't think about its running out someday. And we didn't think about the harm its fumes could do to the atmosphere.

Carbon has been a faithful servant to people. It has kept us warm and given us light. It has helped us draw and write. It has given us industry and technology. It has dazzled us with diamonds and the molecules of life. The versatility of the carbon atom made all this possible. Not only has it been a friend to the other elements—it has been a friend to us. Carbon connects us with the plants and the animals and the earth and the sky.

Glossary

allotrope—a form of an element that comes about because of the arrangement of its atoms. Charcoal, graphite, diamonds, and buckyballs are allotropes of carbon.

atom—a tiny particle that is the smallest piece of an element. Each element has a different kind of atom, and atoms bond together to form molecules.

benzene—a clear liquid organic compound that is used as a motor fuel. Its molecule has six carbon atoms linked in a ring with six hydrogen atoms. It is the simplest of a class of hydrocarbons whose carbon atoms form rings.

buckyball—a ball of 60 carbon atoms bonded together. The full name of the material is buckminsterfullerene. It was named after an engineer and architect, Buckminster Fuller, because the buildings he designed, geodesic domes, have the same structure as buckyballs. Buckyballs are a major part of soot.

carbonate—to add carbon dioxide to a drink to make it bubbly.

carbon dioxide—an invisible gas that is a part of air. Animals exhale it and plants absorb it. It is a compound with one carbon atom and two oxygen atoms in each of its molecules.

chemical reaction—a process in which atoms bond together or break apart. Two or more elements may join to create a compound, or a compound may break down into its elements. Charcoal burning in air, for example, is a chemical reaction in which carbon atoms bond with oxygen atoms to form carbon dioxide.

chemist—a scientist who studies the materials that make up everything in the world and how they react with each other.

compound—a substance made up of two or more elements chemically bonded together.

electron—a tiny particle inside an atom. It has a negative charge and makes it possible for atoms to bond with one another.

element—one of 92 basic substances that make up all the matter in the world. Scientists have made about 18 other elements artificially in laboratories. An element contains only one kind of atom and cannot be broken down into any simpler substance.

ethanol—an alcohol that people drink. It is one of many kinds of alcohols, which are compounds containing carbon, hydrogen, and oxygen. Some alcohols are used as fuel and others as antifreeze.

fructose—the kind of sugar found in fruits and honey. It is a compound containing carbon.

gas—a substance that has no definite shape or volume, but spreads into whatever space is available. It is the form a substance takes when the temperature of its liquid form is raised so high that its molecules rise into the air. The "gas" people put in their cars is something else entirely; it is the liquid *gasoline*, which is sometimes called "gas" for short.

glucose—the simplest kind of sugar. Glucose, found in blood, is a carbon compound, or an organic compound.

graphite—a soft, shiny form of carbon used as pencil lead. The carbon atoms are arranged in sheets, so they slide off easily when you write. Because of this property, graphite is used in many places in industry to lubricate machinery. It is also used in nuclear reactors to absorb neutrons.

hydrocarbon—a compound containing only hydrogen and carbon. Different kinds of hydrocarbons are found in natural gas, coal, and petroleum.

iron oxide—a mineral, also called iron ore, that is a compound containing iron and oxygen. Rust is iron oxide.

matter—anything that takes up space and has weight.

meteorite—a meteor that hits the surface of the Earth. Meteors are chunks of rock traveling through the solar system.

methane—an invisible gas that is a major component of natural gas. The simplest hydrocarbon, it has one carbon atom and four hydrogen atoms in its molecule.

molecule—a group of two or more atoms bonded together to form a compound. They sometimes form in elements too, especially in gases.

natural gas—a gas that rises naturally out of holes in the ground. It contains carbon and hydrogen compounds—mostly methane—and is burned as a fuel.

nucleus—the center of an atom. It contains protons and neutrons.

ore—a mineral, a rocky material that contains a metal or some other valuable substance. Copper ore, for example, is a compound containing copper.

organic compound—a chemical containing carbon. Many organic compounds are found in living things, but plastics and other man-made chemicals are also organic compounds.

oxygen—an invisible gas that animals must breathe to live. It is one of the chemical elements. About 21 percent of air is oxygen.

property—a quality or a behavior that a substance displays.

proton—a particle in the center, or nucleus, of an atom. It has a positive electric charge. The number of protons in an atom determines which element it is.

shell—a place where electrons reside in an atom. In each atom, there are a series of shells, each a different distance from the nucleus. The shell closest to the nucleus can hold two electrons, but most of the other shells can hold eight electrons. The real difference between shells is the amount of energy their electrons have.

soda water—water containing bubbles of carbon dioxide.

smelting—a process in which an ore is heated to release the metal it contains. The heat creates a chemical reaction in which the mineral compound breaks down into its elements. Industries produce copper, tin, and iron by smelting.

sucrose—the white, refined sugar people often add to foods and drinks. It is a carbon compound obtained from sugarcane or sugar beets.

Sources

Brock, William H. *The Norton History of Chemistry*. New York: W.W. Norton & Company, 1992.

Donnet, Jean-Baptiste, and Andries Voet. *Carbon Black: Physics, Chemistry, and Elastomer Reinforcement*. New York: Marcel Dekker Inc., 1976.

Hudson, John. *The History of Chemistry*. New York: Routledge, Chapman, & Hall, 1992.

Mantell, Charles L. *Carbon and Graphite Handbook*. New York: John Wiley & Sons, 1968.

McGowen, Tom. *Chemistry: The Birth of a Science*. New York: Franklin Watts, 1989.

Newton, David E. *The Chemical Elements*. New York: Franklin Watts, 1994.

Urry, Grant. *Elementary Equilibrium Chemistry of Carbon*. New York: John Wiley & Sons, 1989.

Weeks, Mary Elvira. *Discovery of the Elements*. Easton, Pennsylvania: Journal of Chemical Education, 1968.

Index

Italicized page numbers indicate illustrations.

About the Author

Mark D. Uehling has a master's degree in journalism from Columbia University. He has written about technology and the environment for publications such as *Popular Science, Chicago,* and *The New York Times.* He has worked on the staffs of *Newsweek* and *The Sciences,* a publication of the New York Academy of Sciences. He lives in Chicago.